DETOX YOUR BRAIN

The Fundamental Blueprint to Effectively Kill Obsessive-Compulsive Behavior; Simply the Cognitive Therapy to Overcome Overthinking, Depression, Anxiety, OCD, PTSD, and Negative Intrusive Thoughts

Part-1

BY

Henry Fennimore

Contents...3

INTRODUCTION..9

LEARN TO BE YOUR THERAPIST IN SEVEN WEEKS.......13

Types of CBT..29

Advantages of CBT..46

How do you apply CBT in your daily life?................48

WEEK ONE: Outlining and Starting Your Ambitions and Goals...53

Highlighting Your Objective and Setting Target...............56

WEEK TWO: To Move, To Live and Focus on Life. 61

WEEK 3: To Find Out Your Thinking Patterns........67

WEEK FOUR: To Break Negative or Unhealthy Thinking Patterns...69

Seven ways of clearing your mind of negative thinking......80

Here are three more ways to prepare the brain for different thinking:..85

WEEK FIVE: To Control Your Time concerning the Job That You Have on the Ground............................88

Steps in managing time...88

WEEK SIX: To Question Your Fears and Phobias. 93
Tips to Work by Your Fear 102
Others ways to fight your fears 104

Detox Your Brain

© Copyright 2021 - All rights reserved.

The content contained within this book may not be reproduced, duplicated or transmitted without direct written permission from the author or the publisher.

Under no circumstances will any blame or legal responsibility be held against the publisher, or author, for any damages, reparation, or monetary loss due to the information contained within this book. Either directly or indirectly.

Legal Notice:

This book is copyright protected. This book is only for personal use. You cannot amend, distribute, sell, use, quote or paraphrase any part, or the content within this book, without the consent of the author or publisher.

Disclaimer Notice:

Please note the information contained within this document is for educational and entertainment purposes only. All effort has been executed to present accurate, up to date, and reliable, complete information. No warranties of any kind are declared or implied. Readers acknowledge that the author is not engaging in the rendering of legal, financial, medical or professional advice. The content within this book has been derived from various sources. Please consult a licensed professional before attempting any techniques outlined in this book.

By reading this document, the reader agrees that under no circumstances is the author responsible for any losses, direct or indirect, which are incurred as a result of the use of information contained within this document, including, but not limited to, — errors, omissions, or inaccuracies.

Detox Your Brain

Where a legal or qualified guide is required, a person must have the right to participate in the field.

A statement of principle, which is a subcommittee of the American Bar Association, a committee of publishers and Associations and approved. A copy, reproduction or distribution of parts of this text, in electronic or written form, is not permitted.

The recording of this Document is strictly prohibited, and any retention of this text is only with the written permission of the publisher and all Liberties authorised.

The information provided here is correct and reliable, as any lack of attention or other means resulting from the misuse or use of the procedures, procedures or instructions contained therein is the total, and absolute obligation of the user addressed.

The author is not obliged, directly or indirectly, to assume civil or civil liability for any restoration, damage or loss resulting from the data collected here. The respective authors retain all copyrights not kept by the publisher.

The information contained herein is solely and universally available for information purposes. The data is presented without a warranty or promise of any kind.

The trademarks used are without approval, and the patent is issued without the trademark owner's permission or protection.

The logos and labels in this book are the property of the owners themselves and are not associated with

Detox Your Brain

this text.

Detox Your Brain

You see people complaining about depression and its associated impact, but can take time to point out the cause of the depression. The primary focus of eradicating depression is to change the way you think.

You will be introduced in this part of the book to what cognitive behavioural therapy is all about and how it can help manage depression and other negative mental situation.

Firstly, you'll learn how to incorporate it into your daily life, and eventually, you'll find out how the technique will generate long-lasting relief.

Cognitive-behavioural therapy is a therapy which involves modifying the way you think. This focuses on the present alone, ignoring past issues, unlike most therapy.

This helps you to see the better side of the problem. This is widely used to stop depression and anxiety. This is based on the definition of your emotions and makes you know you will be stuck in a vicious loop by harbouring negative thoughts.

CBT has been used as a successful way to treat bipolar disorder, insomnia, phobia, panic, eating disorders, schizophrenia, etc. CBT helps you understand when your attitude and behavior influences a specific problem and gives you the ability to deal with life issues.

When you view a problem negatively and see it as the end of the world, you don't expect a positive outcome from the situation, so you keep having negative thoughts, you begin to feel bad, and you tend to behave in some way.

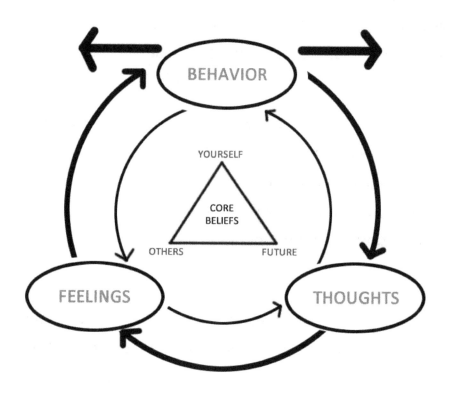

How is this negative thinking going to start? As a child, for example, you have not earned enough attention as other children have, and you begin to see yourself as not being good enough, as being useless and worthless.

This feeling persists, causing low self-esteem and impacting how unusual or peer-friendly you are. This is because pessimistic feelings have been entertained. It becomes a part of you if these negative thoughts go unchallenged or are not thoroughly investigated. You begin to feel weak and insignificant; you start to withdraw from others, and you feel anxious as well.

One of CBT's aims is to become your therapist by learning skills that you can use on your own after the therapy to keep you feeling good, and this part of the book will show you how to be your Seven Weeks therapist. Incredible yes! Let's start now!

Cognitive therapy often helps people look at the "living laws" called schemes. Schemes are abstract constructs or models that organize how you think, feel, behave, connect, and understand and are usually referred to as your personality style.

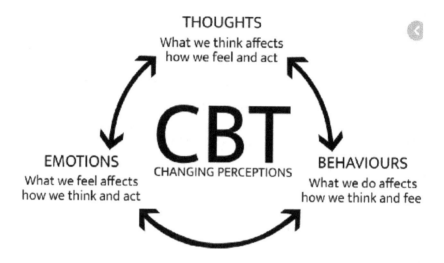

Schemes are beyond conscious awareness and decide how you interpret the world and respond to circumstances.

Although cognitive structures can be adaptive, allowing you to process information rapidly, the same rapid processing that results in maladaptive structures being reinforced in your mind.

This is because they are deep convictions and assumptions about how you should live your life when you grow up.

Collaborative conceptualization (description) is the backbone of cognitive therapy because it incorporates all the underlying concerns so that the challenges are not perceived as a different set of problems.

The therapy can clarify how to maintain current interests and direct therapeutic approaches. Therefore, an initial description of the victim's concerns should direct the treatment plan from the very beginning of therapy to decide which techniques to use.

The successful centred psychotherapeutic approach makes treatment less costly than other forms of treatment, as you can do it by following the instructions on your own. Focusing on skill-building and independence minimizes the course of treatment and reduces the risk of recurrence after treatment has ended.

Cognitive therapists also teach the ability to cope with your life, which has two functions:

1. Allows you to deal with the old issues you've been looking to solve using the Therapy

2. You can use the skills to cope with potential problems in your life.

Essentially, cognitive therapy is a psychotherapy school that recognizes many facets of thought as necessary in the creation and course of emotional and behavioural issues and actualizes cognitive shift capacity. It also helps to progress towards goals and emotional well-being.

Psychiatrist Aaron Beck developed CBT in the 1960s and it can be used for anyone. If you can change a negative thought, it will break all the chains of unfortunate situations.

Cognitive Behaviour Therapy issue and can be used for the following:

1. To alleviate symptoms and problems.

2. To help the person develop skills and techniques to deal with them.

3. To help the client change cognitive processes to prevent a recurrence

Cognitive therapy works with nearly all emotional problems:

Depression

Bipolar Disorder

Shyness and social anxiety

Panic attacks and phobias

Obsessive Compulsions Disorder

Generalized anxiety (chronic worry)

Chronic Fatigue Syndrome

Post-Traumatic Stress Disorder (PTSD)

Eating disorders

Insomnia

Difficulty establishing or staying in relationships

Problems with marriage or other links you're already in

Job, career or school difficulties

Sexual problems

Sexual and physical abuse

Survivors of disaster and trauma

Personality disorders

Childhood and adolescent problems

Aging and elderly problems

Feeling "stressed out" or "stuck."

General Health problems

Low self-esteem (accepting or respecting yourself)

Inadequate coping skills, or ill-chosen methods of coping

Passivity, Procrastination, and "Passive Aggression."

Alcohol and addiction

Multiple Sclerosis (MS)

Shame and Humiliation

Occupational Issues

Anger, Aggression, and Violence

Psychosis and Schizophrenia

Trouble keeping feelings such as sadness, fear, guilt, etc. under control

Hypochondriasis

Over-inhibition of feelings or expression

Irritable Bowel Syndrome (IBS)

Chronic pain

Cognitive illusions are negative thoughts that include negative emotions like:

> i. Filtering: This neglects the positive side and focuses on the negative side of a situation. Jumping into conclusions without thinking properly,Gaving a contrary conclusion without proper consideration of something.
>
> ii. Personalization: This is a misconception where a person assumes that everything, irrespective of how irrational it may be, affects external events or other people.

A person with this distortion may feel that they have an exaggerated part in the negative things that are happening around them. For example, a person might think that arriving at a meeting a few minutes late led to it being interrupted and that if they were on time, everything would have been better.

iii. Blaming: You are responsible for the outcome if events don't go the right way, and you immediately assume.

iv. Always Being Right: This implies being right is more important than the perception of others.

v. Mislabelling: if you fail in a specific task, for example, then you conclude that you are a failure.

vi. Emotional Reasoning: To assume that without proper investigation, the emotional reaction indicates something as right or wrong.

vii. Selective Abstraction: To be biased, to concentrate on a particular part of a description (to recognize it as accurate) and to ignore the other part

viii. Dichotomous Thought: Dichotomous thinking is a symptom of many psychiatric conditions and personality disorders, including borderline personality disorder (BPD). This leads to emotional and behavioural dysfunction and interpersonal issues.

ix. Catastrophe: It involves having irrational thoughts, thinking and to imagine something worse than it really was.

x. Should Statements: Should statements can be a typical negative thinking pattern, or cognitive distortions, which may contribute to feelings of fear and worry.

xi. Over-generalization: Seeing an adverse event as an endless pattern of' always' or' never again.'

The CBT steps are:

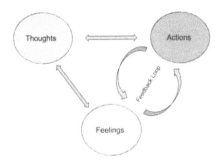

i. Identifying problematic conduct: Pointing out the activity that follows the negative thoughts and bowing the issue is taking a toll. Sit down to examine what the negativities are encouraging. Carefully check every area of your life to illustrate each negativity.

ii. Decide whether habits are excessive or deficient: you decide how behaviours affect your life.

iii. Assessing the essential activity for the length of the frequency: by doing so, you find out how much time you spend making a specific habit — wallowing in self-pity, for example, as a habit you are continually putting on. You find out how much time you spend doing this or surfing the web and digesting irrelevant material. If it's 18 hours a day, you test the amount of time spent on social media.

iv. If it's too much, try to reduce the behaviour: if you know that you spend a lot of time digesting meaningless content on social media, you can cut it from 18 hours to 16 hours, to 14 hours, to 12 hours, you can drastically reduce it.

Types of CBT

1. Brief Cognitive Behavioural Therapy (BCBT): This is a therapy that has been widely used in the use of excessive abuse.

Changes in behavior include understanding different behavior. For you to let go of negative thoughts, you'll need to nurture your mind with positive thoughts. By agreeing and being ready to stop the negative thoughts.

It takes place over a few sessions that can be summed up to 12 hours, and you cannot witness a complete turnaround. David M. Rudd had suggested it to prevent suicide.

Behavioral methods believe that patterns of substance abuse are established and sustained by general learning and reinforcement concepts of your mind.

The early behavioral methods of substance abuse were primarily inspired by the ideas of both classical Pavlovian activity and Skinnerian learning ooertions. Currently, psychological therapy is based mainly on the management of substance abuse problems by approaches drawn from both operative and classical learning theories, though not exclusively.

The leading advantage of behavioral therapy is that it can be changed by altering the situations that control this behavior because substance abuse is a learned behavior pattern.

This goal can be accomplished by either concentrating on the common practice of addiction responses or on the patterns of strengthening the mind.

More precisely, the classically conditioned response can be tackled through either stopping or counterconditioning procedures; the reaction of the operator can be targeted through disaster management or training in coping skills

2. Cognitive Emotional Behavioral Therapy (CEBT): The extended version of CBT that was initially developed by Dr. Emma Gray (nee Corstorphine) for people with eating disorders but has since been used to help people with a broad range of emotional experience and speech problems, including anxiety, depression, obsessive-compulsive disorder (OCD), post-traumatic stress disorder (PTSD),

low self-esteem and fear of harm.

You would benefit a lot from knowing and recognizing that your feelings, emotions, and thoughts do not necessarily reflect reality.

If you allow them to pass without judgment through consciousness, healing can begin. You can prevent being overcome by even strong emotions by learning to recognize and acknowledge negative emotions and feelings.

This allows calmness by your emotions to vanish and reappear.

Problems include anxiety, depression, OCD, post-traumatic stress disorder (PTSD), and anger issues. It incorporates elements of CBT and dialectical behavioral therapy, which aims to bolster cognitive understanding and tolerance to reinforce the therapeutic process. It is often used as a "pre-treatment" to prepare people for longer-term rehabilitation, and better equip them.

Cognitive Emotional Behavioural Therapy's (CEBT) core components include:

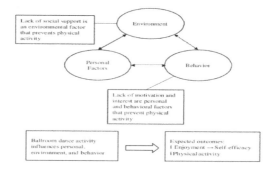

Knowledge about Emotions and Their Functions: Being enlightened about your feelings and the tasks that each emotion plays in a person's life.

Strategies to raise awareness and confidence when coping with emotions: being aware of your feelings and how to deal with them. Understanding when things aren't smooth and how to handle this without hurting you.

Motivation to Change Long-Standing and Unhelpful ways of Dealing with emotions: Aim to improve your attitude when dealing with unhelpful circumstances.

When handling situations, applying new strategies, and not letting go of the good ones — techniques to restructure assumptions about emotional experience and expression: using ideal psychological methods.

3. Structured Cognitive Behavioural Training: This assumes that behavior is inextricably linked to values, thoughts, and emotions that are designed to bring a person to a specific outcome over a particular time. SCBT has been

used to address addictive behavior, particularly with substances such as cigarettes, alcohol, and food, to solve diabetes, and to cope with stress and anxiety.

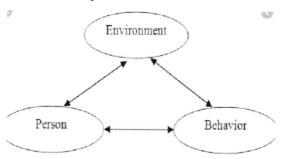

In the field of criminal psychology, SCBT has also been used to minimize recurrence, which is the habit and the want of a convicted criminal to attempt to commit the crime again.

It helps people better manage and control their situations by instilling values that aim to overshadow their negative thought and make them act positively.

4. Moral Resonation Therapy: It is a therapeutic approach that helps to reduce recurrence among the youth and adult offenders by

improving their moral reasoning. This helps to overcome the condition of antisocial personality. The MRT workbook is structured around 16 concrete steps (units) that concentrate on seven fundamental treatment problems:

- The confrontation of attitudes and behaviors of beliefs: questioning and understanding negative attitudes
- Current relationship assessment: Reflect on and analyze the ongoing relationship and recognize where the negative thoughts come from.
- Reinforcement of positive behaviors and habits: You empower yourself with constructive thoughts and behaviors iv when you can identify the negativities and kill them.
- Optimistic identity formation: development of a distinct and positive personality. Personalities emerge not only by clearly understanding what you see, but also by responding to them. The

character describes how you can handle or enter a situation.

- Self-concept Enhancement: This means a positive perception of yourself. It is the inspiration that will make you feel good about yourself and will increase your self-esteem. By doing so, you're discovering yourself (trying out new things), stop the comparison of yourself to others (you're special, you're not everybody else), you'll be able to escape your self-image by not stagnating, keep updating your self. Don't be afraid to make mistakes because you're going to screw up a lot of times before you do it right. You set yourself realistic goals and should not stop working until you reach them.
- Decrease in Hedonism and Anger Tolerance development: According to the Cambridge dictionary, hedonism is living and behaving in ways that ensure you have as much fun out of life as possible, believing that the real important thing in

life is enjoying yourself.

The fact that you have to lead only one experience does not mean you have to live a wasteful one. You are supposed to live a productive life. Too much of all is considered a bad thing, so everything in life should be assessed appropriately.

You ought to get rid of hedonism. No one tells you shouldn't enjoy life, but you should be wise. Create tolerance for frustration; not all failure can come into you. Life will test you with challenges, and if you continue to give in all disappointment, it would be considered terrible for your mental health.

You should learn to cope with stressful situations. You will succeed if things don't go the right way. Begin your day with a positive note, a positive mindset, and engage your thoughts with funny things that help to reduce stress.

Whatever may have been bad for your day, don't give up. Positive affirmations end your day. Don't give up the situation due to the frustrations.

- Higher Stage Development of Moral Reasoning: You can differentiate between what is right and what is wrong. It's a necessary process that people use when trying to do the right thing, and often every day. Every day, for example, people face the issue of whether or not to lie in a given situation.

People make this decision by reasoning the value of their potential actions and by balancing against the possible consequences of their actions. From the wrong thing, you should be able to identify the right thing.

It's imperative because it gives you strategies to decide what it's good to consider and what it's terrible to dismiss.

5. Stress Inoculation Training: This uses a mixture of cognitive, mental, and some methods of humanistic psychology to reach the victim's stressors.

Pain Inoculation Therapy (SIT) is a form of psychotherapy that seeks to allow patients to train themselves in advance way ,to effectively manage stressful events with minimal distress or fuss.

This is widely used after stressful circumstances to help people cope with their stress or anxiety. SIT's first stage is a therapist that illuminates the patient about the essence of fear.

The belief that, through the unconscious process of poor coping patterns, people often and instead unconsciously make their stress worsen.

Ultimately, before moving on to the next step, the therapist aims to establish a better understanding of the innate of stress elements faced by the individual. The second step is about learning skills in stress management and putting them into action.

The specific choice of skills taught is essential and needs to be tailored independently to the wishes and abilities of patients and weaknesses to become effective.

Depending on the particular needs of the patient, a range of emotional regulation, relaxation, cognitive assessment, problem-solving, communication, and socialization skills can be selected and taught.

The final phase includes applying the learned skills to daily tasks before you can efficiently handle stress.
6. Mindfulness-based cognitive behavioral therapy: To help individuals better understand and control their thoughts and emotions. It explores subconscious patterns to achieve relief from feelings of distress.

You will understand how to use cognitive techniques and practice of mindfulness in this counseling approach to address the repetitive cycles that often cause depression and other mental health problems.

During situations of depression, low mood, negative thoughts, and specific symptoms of the body such as weariness and sluggishness frequently occur together. MBCT helps you to understand how to identify your sense of well-being and make yourself look separate from your thoughts and moods.

It gives you the ideal tools to destroy negative thoughts as they arise and depressive moods. MBCT promotes good health and assists with anxiety and depression treatment.

7. Unified Protocol: It can be used for many anxiety and depression conditions combining elements such as concentration, cognitive therapy, and behavioral therapy.
8. Rational Emotive Behavioural Therapy: Rational Emotive Behaviour Therapy (REBT) It is a short-term form of psychotherapy that allows you to point out self-defeating thoughts and feelings,

challenge the rationality of these feelings and replace them with healthier, more productive, more positive thoughts and beliefs.

Most of the time, REBT focuses on helping you understand how unhealthy thoughts and feelings create emotional distress, which in turn leads to harmful actions and behaviors that interfere with your current life goals.

Once identified and understood, negative thoughts and activities can be changed and replaced by more positive and productive practices, enabling you to develop better personal and professional relationships.

REBT can help you with negative emotions such as anxiety, depression, shame, and extreme or inappropriate frustration.

It also helps to improve negative and self-defeating habits such as violence, unhealthy eating, and procrastination that obstruct your essence of life and can stop you from achieving your goals.

9. Self - instructional Training: The therapy will help you to control your behavior. This is used to change faulty belief. "Everybody hates me," for starters.

This belief has an impact on your life actions and attitude. You're back on the assumption that nobody loves you, and no matter how far you go, you'd never be loved.

This harmful ideology is going to affect your life and ruin your self-esteem. You are antisocial because you don't want to go out to meet new people or make friends. You are more likely to be constrained.

If not questioned, the way you react with people will be affected. Mostly, the therapy is used to get rid of such values and adopt positive ones.

Advantages of CBT

- It helps to sort out negative thoughts and emotions.
- It helps to control stress and anxiety disorders
- It helps to handle relationship issues.

- It encourages you to deal with grief and loss.
- It is very successful as a medication to treat many mental health problems: about 7% of Grownups struggling with depression in the U.S each year and the main symptom change their thinking style when CBT is implemented.
- Unlike other types of talk therapy, CBT does not take much time to complete: as stated in the heading, only seven weeks is enough for you to pursue CBT.
- CBT contains series of forms and methods of therapy sessions.
- CBT results in rapid change. You begin to see results as quickly as possible.
- There is no age limit, as it operates with adults and young people alike. It works for someone else.

How do you apply CBT in your daily life?

The best thing to use in daily life is to start small and be mindful of the circumstances you are trying to change.

Be mindful of destructive emotions, toxic thinking that involves generalizing, blaming yourself, wallowing in contempt for yourself. All of them can be called cognitive illusions. It's been listed above.

You begin to have power over your thoughts when you are aware of these thoughts and can control them.

Reframing your patterns of thinking: What do you do when you find that you are engaged in negative thoughts? You improve them if you notice that the belief appears harmful to you. This will help to change your perception.

You should be able to break down problems into smaller bits and address them little by little. How are you preparing for a solution to the issues, if you view them negatively? Essentially, this is what CBT is all about. Your feelings, as they are interconnected, affect your actions towards a situation.

Running the tape through this method is particularly useful for men and women to heal the froms of addictions to drugs. Symptoms of cognitive distortions are cravings and intrusive thoughts about.

Imagine what's going to happen if you relapse and fall back to your negative thoughts. Find reality as opposed to fiction. Relapse is never worth it at the end of the day. Failure is never an option.

Now that you know the definition of CBT, the forms, the advantages, and how it can be applied to your everyday life. For seven weeks, you'll learn how to put them into action.

The below table lays out a plan of seven weeks to become your therapist with a strategy to be used in situations management every week.

WEEK ONE	Outlining and starting your ambitions and goals.
WEEK TWO	To move to live and focus on life
WEEK THREE	To find out your thinking patterns
WEEK FOUR	To break negative or unhealthy thinking pattern
WEEK FIVE	To control your time concerning the job that you have on the ground
WEEK SIX	To question your fears and phobias
WEEK SEVEN	To put them into practice.

WEEK ONE: Outlining and starting your ambitions and goals.

WEEK TWO: To move to live and focus on life

WEEK THREE: To find out your thinking patterns

WEEK FOUR: To break negative or unhealthy thinking pattern

WEEK FIVE: To control your time concerning the job that you have on the ground

WEEK SIX: To question your fears and phobias

WEEK SEVEN: To put them into practice.

The above table lays out a plan of seven weeks to become your therapist with a strategy to be used in situations management every week.

Detox Your Brain

WEEK ONE: Outlining and Starting Your Ambitions and Goals

The first thing you do is set a goal/objective for yourself when using Cognitive Behavioural Therapy. "I need to be capable of doing this at the end of the seven-week therapy, and I wish I could do that..."

At just the finishing of the treatment session, you make a list of what you want to have achieved, and then you strive to get it through. After all, you can't be afraid of something and expect it to go where you want without getting it done.

You're looking for what you want and making extra efforts to get the results you're hoping for. Depression:

The causes of depression are so many. When you feel that you are not good enough when you think that you do not fit in, you feel bad or left out, and you believe that people dislike you, these feelings cause depression.

Certain mental illnesses, such as erratic feeding and symptoms of frustration. You set your goals; "I aim to manage my rage at the end of the therapy for seven weeks, I want to be able to regulate how I eat at the end of the therapy for seven weeks." If the mood is shifting (alternating between a jovial mood and a depressed mood).

You should be able to stabilize the situation at the end of the therapy. All mental illnesses should be addressed, and you should figure out how the best way you can do that at the end of the treatment. You move on to the next cycle the next week after deliberately doing so for the first week of the counselling phase.

Highlighting Your Objective and Setting Target

The fundamental step to making that dream come true is firstly making it an idea. A person who wants to become a neurosurgeon must make his first step by planning to go to medical school. You need to set your goals, plan and figure out ways to achieve them.

There are guidelines for achieving your goals, first off:

- Be wise about your dreams
- Be involved in your dreams

- Be reasonable in your dreams
- Meet experts

Be wise

Your vision would make you a better person in society. What societal value does your goal contribute to? Ask yourself what people would benefit from your dream, if it were helpful or harmful.

Many people who have visions of becoming criminals, and thugs are not aspirations that society would be proud of. A doctor's capacity for society's development is more important and useful than a thug's.

Expecting Obstacles

You don't have anything with you. You need to fight for everything you get. To achieve these goals, there are particular challenges you'd face when you're pursuing or trying to reach your target, there are different factors that would influence the individual's achievement of the goal, and they include

- Environmental factor
- Psychological factor
- Physiological factor

The environmental factor is an individual's relationship with his environment and how it influences his or her decisions.

An entrepreneur would find it difficult to survive in a land that doesn't have the sufficient facilities that are intended to enhance his / her business, and a business development will significantly decide how environmentally friendly a new company will suffer from unfavorable conditions.

Psychological Factor: These are factors that affect an individual's mental decisions and include the way an individual looks at life. Pychologically we are all different, we think and react differently to situations.

Examples of physiological factors are isolation, depression,happiness, to name a few.These factors may create either a psychopath from society or someone who is a gift to the community

Detox Your Brain

WEEK TWO: To Move, To Live and Focus on Life.

Look at your life. "How am I living my life?" "How do I handle circumstances around me?Identify your most important values and responsibilities. Identify why you did it right and how you lagged.

Take time to figure out how much time you spend on daily tasks and how it affects you negatively or positively.For example, you may know that you spend 3 hours browsing every day. If this behavior is not highly valuable to you, try to schedule 1 hour per day for these activities so you can free up to 2 hours per day for more important things.

Reflect on how you have caused your everyday life to be influenced by your thoughts and feelings. On a wrong note, you start the day, harbor negative thoughts, and how it affects your day.

Will you spend time wallowing in self-pity, criticizing yourself for wasting your day?

Do you spend your day filtering evil thoughts out of the good ones and devoting yourself to the evil thoughts? Every of these should be researched carefully.

Going to cite two examples to justify this story.

A man named John was almost a perfect man, he had it all for him, he married a beautiful wife.

At the beginning of their marriage, things flourished, he had a good-paying job, with two beautiful children and had the world at his feet while working as a marketing manager in a business.

One day he lost his job, he couldn't feed his family, so his wife moved out, and he couldn't provide the means to take a care of his children.

He was devastated and unhappy ,he had never faced this type of situation before and he didn't know his way around it, but one thing was sure; he didn't give up and let himself down.

He later got a job as a salesman, he earned stipends as wages, but he managed to make his way.He continued to make money for his employer, and his boss wanted to hire him on a long term with an excellent package.

He continued working hard and bagged promotions after promotions, and eventually, he became the head of sales.

One fateful night, he worked so late and decided to go to the pub to have a drink before heading home, while drinking he saw a very familiar face, he decided to go over to know who it was.

When he was immediately up close, he suddenly recognized the person, his long-term colleague, who also lost his job at the same time in the company they had worked together.

Wanting to know why his old colleague was doing so poorly, his friend began to describe how life was unfair to him, he said, he lost everything right after his job and began to drink and play every night to detach his mind from the reality that he was a loser.

The take-home message is that while John suffered from hardship and unfavorable situation of his sack, he did not give up on life. That's what life means.

Of course, life can beat you down a lot of times, it's natural to get beaten down, it's how you recover that decides you.

Some individuals tend to see the cup half empty while some people see it half full, the most crucial thing is to get the cup filled back to it. The ability to be positive in all situations is a rare good gift.

You ought to know that life is fun, so you ought not to be too harsh on yourself. Sure, things are happening, but you have to keep going.

WEEK 3: To Find Out Your Thinking Patterns.

You will recognize the pattern of your thinking. How are you going to reason? Recognize the negative thoughts and the impact your reaction to your situation.

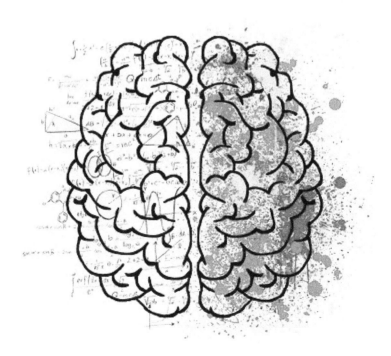

A way of thinking due to thinking in a specific pattern of learned behavior.

Better thinkers, for example tend to identify and analyze problems, test hypotheses, explore explanations, and be analytical. Third week is to identify that negative thinking before heading to the next step.

A negative thought is a behavioral habit. This feelings sink in and stay and to get rid of them can be challenging for some people.

It can be tempting to try to drive those feelings out of your mind when you first start worrying negatively. You want to stop thinking about them and drive them out as fast as you can.

This strategy can backfire, though. Suppressing those negative thoughts can reinforce the pattern of thinking and make things worse.

The harder you try not to dwell your mind on something, the more it keeps staying around it.

You need to try a different approach to remove negative thinking, something that will once and for all rid the mind of those negative thoughts.

WEEK FOUR: To Break Negative or Unhealthy Thinking Patterns.

Depressive thought is a way of thinking in which people strive to see the worst of all or minimize their aspirations by taking into consideration the worst possible scenarios.

The alternative way will be optimistic thinking, with a constructive outlook addressing situations or circumstances.

You may be distressed by the common cold, fatigue, stress, hunger, sleep deprivation, even allergies, which contributes to depressive thoughts.

Typically, these interpretations are used to reinforce negative thinking or emotions.

Depression can be caused in many situations by negative thought itself.

When you think a lot about negative thoughts, you will most likely deal with negative emotions such as frustration, depression, apathy, anxiety, fear, and more. Sometimes you might even feel very helpless.

One of the significant challenges you can face while dealing with these situations is overcoming negative thinking. After all, while you are harnessing all these fantastic new tools that help you think positively and look to a brighter future, you are still struggling with unhelpful restricting beliefs from the beginning of your life.

Many of these beliefs can go uninhibited and begin to disrupt your vision of a better life.

A lot of sad thoughts through your mind can be stressful because you can't be sure whether depression will make you think negatively, or you will be depressed by thought negatively.

You may be distressed by the common cold, fatigue, stress, hunger, and sleep deprivation, even allergies, which contributes to negative thoughts.

Luckily, however, there are a lot of practical things you can do to help you break harmful patterns of thinking. Here are five of the easiest ways of stopping negative thinking.

1. Thought Stopping:

When you notice that your mind is beginning to enter negative thoughts or images, try to say, "stop!" "to yourself.

If you're alone, you can try to answer this loudly, but when you say it in your head, it can also be hushed.

If you prefer, you can use language that is more powerful than" stop "(such as" Get out of my head! "or something a little more colorful).

Images can be more potent for people who are not as moved by words. The typical representation is the bright red stop mark can see in your mind when intrusive thoughts start to appear.

There are also a few more straightforward approaches to avoiding these thoughts. You may try the old trick of splashing your face with water, for example or change your thought path. Many people like to count from 100 to 1.

2. Positive Statements

Positive words can be used in a variety of ways. First, they could be used in the same way as strategies of being positive to stop thinking negatively.

In other words, as soon as you feel a negative thought coming your way, you might say an affirmation.

For example, if you try to find a new partner and think maybe you don't deserve love, you can say, "I'm a beautiful, lovable person, and I'm going to find a great relationship."

Second, though, saying affirmations daily helps to reshape the mind, making them a powerful tool even if you're in a good mood already.

Carefully construct your claims and try to make eye contact in the mirror as you recite them.

3. Strengthening borders

If you have been around for a long period of time on negative thinking, you may think it's unrealistic to expect yourself to change your mindset suddenly.

Also, comments and strategies to avoid thought may seem to prolong negative thinking for a later date in this case.

Whether this sounds right once it gets to it negative thinking, you might want to spend at least a few weeks establishing boundaries.

The idea here is to choose a set, limited time to encourage your mind to entertain negative thoughts and to actively stop or combat them at any other time of the week.

You will find they seem less healthy and have less ability to control your mind when you are told that you will have time to consider these thoughts.

However, many people find that when they arrive at their scheduled time to allow reflection of negative thoughts, they can't even think of anything and that this helps them break their cycle.

4. Writing and Destroying:

If your negative feelings are correlated with particularly strong emotions such as anxiety, rage, or envy, find a way to let them all out in writing.

Use a pen and paper to show all the pent-up frustration. Then you can choose a way to destroy this paper, which symbolizes your commitment to move forward. You might, for instance, tear it up, crush it into a ball, burn it, or scribble it.

Those who are not as keen to use words to express themselves can have a similar impact on artistic endeavors. You might, for instance, sculpt or paint a representation of your negativity and then kill it (or change its shape).

The point of this strategy is to get some physical description of your negativity so you can banish it symbolically.

One of the significant challenges you might face when coping with life is to overcome negative thinking.

After all, while you're harnessing all these fantastic new tools that help you think positively and look to a brighter future, you're still grappling with unhelpful restricting values from the beginning of your life.

5. Reason with Yourself

When you believe you are beginning to descend into depression, you can also try to reason with yourself.

This technique involves identifying an expression that you can recite to yourself to remember that you have control over your body responses and increasing that power over time.

Practice this method by taking a deep breath and saying something like "Just because I've had some bad relationships doesn't imply i need to do that to my body" or "Just because I've been struggling to find a good job doesn't mean I'll never find one in the future."

Say "Relax now" (letting the word "relax" be your cue to exhale, letting out stress and negativity) after the chosen phrase.

Seven ways of clearing your mind of negative thinking.

1. **Changing The Style of Your Body**: Take a moment to consider the style of your body. Were you slouching with a closed position? Are you frowning? If you are, you will have a better risk of thinking negatively. Bad state of the body can reduce your self-image and contribute to a lack of self-trust.

It's only natural to start having evil thoughts in that emotional state. In a relaxed manner, sit up straight. Open your spot and smile more.

Fix the posture of your body, and you will feel much better. Maybe it is just that you need to clear these negative thoughts.

2. **Work It Out**: Negative thinking often happen when you have issues or feelings that you need to let out about your situation. Keeping things to yourself is not good.

When you have something to deal with, you should work with someone about them. Putting things into words will help shape your thoughts. This might assist you to put things into track so that at the root of the problem, you can deal with them.

3. Spend a minute Relaxing all Your Emotions: As your mind races , it could be challenging to keep up.

   For all things running through your head, the internal thoughts – particularly the negative ones – can be challenging to control.

   Sometimes it takes only a minute to calm down. It's like meditation; you make up your mind. Treat it as a reboot. You can fill it up with something a little more optimistic.
4. Adjust the tone of your thoughts: Often, the result of poor perception is negative thinking. Look at the point of view that you are having on the issues that are happening around you.

   Instead of reasoning, for example, "I'm going through a hard time, and I'm having trouble," think, "I'm facing some challenges, but I'm working on finding solutions." You're just repeating the same thing, except that the second approach is a more optimistic .

But this slight tonal change can often make a big difference to your habits of thinking.

5. Be imaginative: This will pay to spend some time creatively when negative thoughts come. Explore your emotions with a creative outlet. Write out the stuff. And if you have to ,use a crayon, draw or paint something. It can work as long as you use your imagination to break out of your negativity.

   Exploring your feelings into arts acts like auto-therapy can boost your mood. It can feel like a release of negativity. You get them out of your mind when you put your feelings through an image form and flush them out.

6. Take a walk: If emotions pop up in the head, it's safe to believe they're produced there. Okay, that's accurate, just in part. Our thoughts are often a product of our surroundings. For instance, if you surrounded yourself with stressful people and negative imagery, in turn,

you would probably start thinking negatively.

This can help tremendously to get away from a negative environment. Get away from your usual atmosphere by walking alone — brain uplifting like a park or a museum somewhere.

Time spent distancing yourself from these negative influences will bring great peace of mind to you.
7. Start Concentrating on What Makes You Happy :Even we lose sight of all the ways things go right about our daily lives .

If that's you, maybe you need to re-train your mind to concentrate on all the right things happening around you rather than the wrong things.

List everything you're grateful for, no matter how low they look. Don't take it for granted anymore. The good ones in our lives are even right in front of our faces, and we somehow don't see them. Avoid being blind to the good things you've always done for yourself or happened to you.

Here are three more ways to prepare the brain for different thinking:

1. Reorganize Things That are Not Helpful: It's not beneficial to think stuffs like "This can't be true," or "I'm such a fool. I've destroyed it all." Weak assumptions tend to turn into self-fulfilling prophecies. And overly negative feelings are stopping you from taking affirmative action. But the good news is, with

more thoughtful comments, you can respond to unhelpful thoughts.

When you think, "No one will ever employ me," note, "If I keep working hard to look for jobs, I may improve my chances of being hired." Or, when you think, "This will be a failure," look for evidence that your attempts may be successful.

So, make a more objective statement like, "There's a risk that this won't work out, but there's also a chance that I could be successful. All I can do is my best."

2. Make a mistake: Sometimes, your mind is lying to you. So, consider it a challenge when it tells you that you may not be able to get a promotion or that you will never be able to complete the job you're working on.

After you think you're too tired to keep going, force yourself to take another step. Or challenge yourself to continue applying for promotions despite the insistence of your brain that you're not going to get a new position.

You must train your brain to see yourself in a different way every time you successfully prove your negative predictions wrong. Over time, in a more accurate view, your brain will begin to see your limitations as well as your capabilities.

3. Make your mantra: Take stock of your habits of negative thinking. Do you call the names of yourself? Or are you thinking about doing stuff you might forget to do?

Then create a personal mantra that you can use to communicate with the negative messages. Repeating things such as "Make it happen" and "Do your best" filters the negative out. And over time, you're going to grow to believe these claims more than the unhealthy things that you informed yourself.

WEEK FIVE: To Control Your Time concerning the Job That You Have on the

Ground

Time!!!... The most expensive asset in life. Time is the difference between success and failure. Time cannot be recovered when wasted foolishly. As we all know, time waits for no one.

A young boy was admitted to a university at a tender age of 16, and he was hell-bent on finishing at age 20, he set his eyes on achieving his goal, after four glorious years at the university.

But on the verge of completing his school he decided to procrastinate due to some outside work until there was not enough time to study .He later had an extra year . You should know the wisdom of using your time.

Steps in managing time

- Be proactive
- Set timely reminders
- Be disciplined

Be proactive:

The proper adjective describe a person actively doing things. If you're optimistic, instead of waiting for things to happen to you, you'll start working on the situation.

And, if you're cautious enough, you'll be ready before something happens. The opposite is to be reactive or wait for things to happen before you respond.

Speaking of the cold season in the winter. A proactive person will wash his hands and take vitamins; a reactive person would get sick and have to take medicine for a cure.

Being active will help a lot in every scene; you should be able to act and build plans to help solve problems. If you are dealing with these circumstances in real-time, this will help.

Set timely reminders:

Create a list of what you want to do in a day and make sure they are done daily. Doing this lets you allocate time concerning your work and avoid quite so much time spent on a specific task to the exclusion of other tasks.

Be disciplined

Do not go to bed the same way you woke up, be diligent enough to follow your dreams, and make sure you achieve it.

Procrastination is the greatest obstacle to time, and the desire to be self-disciplined when you set your sight to a target is indispensable. Do what you can do as from today.

WEEK SIX: To Question Your Fears and Phobias

What do you worry most about, what is that one thing that sends your spine shivering down?

As humans, we are afraid of many things; some are afraid to make decisions because they don't want the public to look down on them. The biggest dream killer is "WHAT WILL PEOPLE SAY."

What's the fear? Fear is, in my experience, an uncomfortable feeling caused by the perception in danger, real or imagined.

Anxiety could be either actual, or an action played on us by the subconscious following this concept. Fear has led to many dreams going to their graves early.

The first step to confronting your phobia is to understand them first, where they come from.

Whether they are a figment of your imagination, or whether they are real, you can start slowly after doing this, you don't have to face them all at once. It takes one step at a time to climb a ladder.

An example, a little girl was once afraid of the dark, and she would run to her mother to tell her that she couldn't sleep. With two choices left to her parents;support her, or just let her be, fight her fear.

Her mother decided to help her battle it, and her room was illuminated with four lights, at first, her mother would turn off a view that would make a small portion of the room night after the girl got used to it, her mother would switch off two.

Now a more significant part of the room was dark, and the mum persisted until the girl was finally able to get used to the dark. A man who wants to become a skydiver shouldn't be afraid of height.

Furthermore, one that needs to be successful shouldn't be afraid of the hard work, endurance, and other factors that come with it. Remember that phobias is not a sentence of death.

Do not feel less than the next man because you feel more frightened of him because you have the will to conquer your fear.

Many people have abandoned their aspirations due to fear, while others have used their fear to step beyond what others expected.

No one is born of fear. Therefore, concerns can be stopped by the regular application of self-discipline until it stops. Fear of failure, deprivation, and loss of money are the most common fears we face, which often destroy all hopes for success.

Such worries lead people to avoid any danger and reject opportunities when presented to them.

We are so afraid of failure that when it comes to taking any risks at all, we are almost frozen. Many other fears impact our happiness.

People are afraid of love loss. People are afraid of job losses and financial security. People are afraid of embarrassment or mockery.

People are afraid of rejection and criticism. People are afraid of others' loss of respect or esteem. Throughout life, these and many other fears that can hold us back.

In a situation of doubt, the most common reaction is the attitude of "I can't!" This is the fear of failure, which prevents us from acting. It is physically felt, starting in your stomach's pit.

If people are terrified, their mouth and throat are dry, and their heart starts to pound. We breathe shallowly at times and our stomachs churns.

These are all physical manifestations of the pattern of inhibitive lousy behavior that we all undergo from time to time.

Fear Shuts Our Brain Down

This sometimes shuts down the brain and causes the person to return to the response of "fight-or-flight." Fear is a horrible emotion that threatens our joy, and throughout our lives will hold us back.

Visualize Yourself as Unafraid

By seeing yourself acting in an environment where you are not fearful of trust and integrity, your visual image will ultimately be recognized as guidance for your success by your subconscious mind.

Your self-image, the way you see yourself and think of yourself, is eventually altered by feeding your account to the best of these positive mental representations of yourself.

Practice Acting "As If"

By using the "act as if" form, you are walking, communicating, and carrying yourself precisely as you would if, in a particular situation, you were unafraid.

You are standing up straight, smiling, and moving fast and confident, and acting as if you already had the confidence you need.

Use the Law of Reversibility

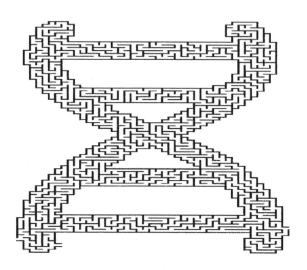

The Law of Reversibility states, "When you feel something, you will act in a way that is compatible with that feeling."

But if you act in a way that is consistent with that feeling, even if you don't feel it, the Law of Reversibility will establish the atmosphere that is consistent with your acts.

Confront Your Fears Immediately

The secret to happiness and success is your ability to confront, cope with, and act through the fears.

You can quickly act to identify a person or circumstance in the life you are afraid of and commit to deal immediately with this situation of fear.

Do not allow another minute to make you sad. Commit to facing the situation or person and put the fear behind you.

Move Toward the Fear

This becomes smaller and more manageable as you recognize fear and force yourself to move towards it. Moreover, when your worries are growing shorter, your confidence is growing.

Your problems will soon lose control over you. In contrast, if you go back from a situation or entity that triggers fear, your fear is growing bigger and bigger.

It soon consumes the thoughts and feelings, scares you throughout the day, and may even keep you awake at night.

Deal with the Fear Directly

The only way to address fear is by confronting it head-on. Most people's natural inclination is to deny that they have a problem caused by some fear. They're afraid to face this. It becomes, in effect, a significant source of tension, unhappiness, and psychosomatic illness.

Be prepared to deal directly with the situation or individual. As William Shakespeare said, "Take arms against a sea of trouble and put an end to them."

If you force yourself to face every circumstance in your life that causes fear, your self-esteem increases, your self-respect increases, and your sense of personal pride increases, you would now get to the stage that you won't be afraid of anything in life again.

Tips to Work by Your Fear

Allow yourself to sit 2-3 minutes at a time with your anxiety. Breathe in and say, "It's all right.

It feels crappy, but feelings are like the ocean— the waves are ebbing and flowing." Have something scheduled to cultivate right after your 2-3-minute sitting time is over:

Call the good friend waiting to hear from you; immerse yourself in an activity that you know is fun and stimulating.

Write down the things for which you are grateful. If you think you're in the wrong place, look at the list. Include it into the list.

Remember that your anxiety is a wisdom storehouse. Write a letter, "I'm no longer intimidated by you, dear Anxiety. What are you going to teach me?"

Exercise. Exercise can refocus you (only one thing at a time can your mind focusses on).

Whether you're going for a short walk, going to a boxing gym for an all-out sweat session, or turning on a 15-minute yoga video at home, exercise is right for you and helps you feel better.

To deflate your worst fears, use humor. What are some ridiculous worst-case scenarios?For example, What could happen if you accept an invitation to address an audience of 500plus people?

I could piss at the microphone in my underwear. I'll be arrested for making the worst speech in historyMy first boyfriend (girlfriend) will be in the crowd and heckle me.

Acknowledge your courage. During difficult times, I'd say to myself, "Every time I don't let fear stop me from doing something that scares me, I'm getting stronger and less likely to let the next fear stop me."

Others ways to fight your fears

1. Take your time off: When you're overwhelmed with fear or anxiety, it's impossible to think clearly. The fundamental thing you ought to do is take some time out so you can calm down mentally.

    Distract by walking around the road, making a cup of tea, or having a bath for 15 minutes from the concern.
2. Breathe through adrenaline: The best thing is not to fight it if you start getting a fast pulse or sweaty palms. Remain where you are, and feel the tension without disturbing yourself. Place your hand palm on your stomach and slowly

and deeply breathe.

The objective is to make the mind become accustomed to the panic that takes away the our fear. Use this form of relaxation for depression.

3. Face your fears: Avoiding feelings makes them more frightening. Whatever the concern, it should begin to fade if you face it. For example, if you're panicking into a lift one day, the next day, it's best to get back into a lift.
4. Think the worst: Try to imagine the worst thing that can happen– maybe you will panic and have a heart attack. And try to think about a heart attack. It is simply not possible. The more you chase it, the more fear will run away
5. Look at the Evidence: It can help question negative thoughts sometimes. For instance, if you're afraid of being stuck in a lift and suffocating, ask yourself if you've ever heard of somebody happening to this. Ask yourself what you'd say to a like-minded friend.

6. Seek not to be perfect: Life is full of tension, yet many of us believe our lives need to be flawless. There will always be bad days and failures, and it is essential to remember that life is messy.
7. Visualize A Happy Place: Pause for a moment to close your eyes and envision a safe and quiet environment.

   It might be an image of you walking on a beautiful beach, or snuggled up with the cat next to you in bed, or a happy childhood memory. Let the positive feelings calm you down before you feel more comfortable.
8. Think About It: Expressing anxiety takes away a great deal of it. It relieves the pain.
9. Go Back To Basics: To self-treat anxiety, many people turn to alcohol or drugs, but that will only make things worse. Easy daily solutions such as a good night's sleep, a healthy meal, and a walk are often the best anxiety cures.

10. Reward Yourself: At last, have a treat for yourself. For example, if you've made that call, affirm your success by treating yourself to a massage, a country walk, a meal out, a book, a DVD, or whatever little gift you're happy with.

CPSIA information can be obtained
at www.ICGtesting.com
Printed in the USA
LVHW081817010621
689062LV00015B/1833